Forestry Commission

Field Book 15

Weed Control in Christmas Tree Plantations

by Ian Willoughby

*Forestry Commission
Research Agency*

and Colin Palmer

*Rural Services
Woodview Cottage
Raycombe Lane
Coddington
Ledbury
Herefordshire
HR8 1JH*

LONDON:
THE STATIONERY OFFICE

ISBN 0 11 710340 3
FDC 414:441:236.1:307:281

Keywords: *Herbicides, Farm forestry, Christmas trees*

Abstract

Good weed control is essential for quality Christmas tree production. Herbicides are usually the most cost-effective way of achieving this. Two principal categories of Christmas tree production can be identified: production in forest plantations and specialist horticultural production. For these two situations, suitable candidate herbicides are identified, and guidance on their safe and effective use is offered.

Disclaimer

Enquiries relating to this publication should be addressed to:

The Research Communications Officer
Forestry Commission Research Agency
Alice Holt Lodge, Wrecclesham
Farnham, Surrey GU10 4LH

Contents

Acknowledgements

John Morgan collected and made available useful information on the subject. Mike Lofthouse, Paul Tabbush and Professor Julian Evans made helpful comments on the text. Typing was by Pam Wright.

Front cover: Two-year-old Norway spruce Christmas trees in a well-wooded area on Yattendon Estate, Berkshire. (41642)

1 Introduction

1

This publication has been written for three reasons:

- good Christmas trees must possess good shape and healthy foliage – this requires, amongst other things, a high standard of weed control;

- Christmas trees are valuable crops and any risk of damage to foliage must be minimized;

- in many situations Christmas tree production is a horticultural activity where special rules apply. These extend the range of herbicides that may be used compared with normal forestry situations.

In common with more conventional tree plantings, weeds compete for moisture, nutrients and light with Christmas trees. Such competition reduces tree growth and survival, and increases the length of time required to produce a merchantable product. Weeds can also detrimentally affect Christmas trees by abrading lower branches, contributing towards the browning and dropping of lower needles, and causing ragged crown growth. All of these traits should be avoided if the intention is to produce a top quality Christmas tree attracting the best sale price. Therefore within Christmas tree plantations, weed control will normally need to take place to a higher standard than in traditional plantations – often 100% weed control for the full 5–8 year rotation is required.

Few herbicide products have specific recommendations for use on Christmas trees. The advice in this publication is based upon experience and trials using herbicides on spruces, pines, and firs, assumed to be the genera of conifers typically grown as Christmas trees. Trials are unlikely to have taken place on the more specialist Christmas tree varieties which are not commonly used for timber production. When considering crop tolerance, field workers should use this publication as a guide to practice, then **test small areas to determine the tolerance of specific conifer species to individual herbicides, before engaging in large-scale treatments.**

2 Types of weed control

2

The use of herbicides is usually the most cost-effective means of weed control. Alternatives include the use of plastic or organic mulches, cultivation, or the use of less competitive ground cover species, but all of these have their own problems. Mulching effectively prevents weeds establishing, but is expensive – the cheapest effective individual plastic mulches cost around 38p per tree or £950/ha at a stocking of 2500 trees/ha (Deboys, 1994), and are only likely to give effective control for 2–3 years. A continuous sheet mulch is likely to cost around 67–95p per tree, although this would probably reduce to 44p per tree at 10 000 trees/ha (G. Drake-Brockman, personal communication). An effective herbicide regime may cost £50–£250/ha/year, with the majority of the costs being incurred in years 2 and 3. However, any mulch effectively eliminates the ever-present danger of damage to foliage from herbicide applications. Cultivation by itself is unlikely to offer sustained weed control unless it is repeated regularly, which may be a costly and technically difficult option. However, a pre-plant cultivation could be considered to aid mechanical planting, to improve early tree growth and survival, and to provide suitable conditions for soil-acting herbicides to work effectively at minimum rates.

Planting trees through a sown grass sward and then simply mowing grass between the rows with no weed-free bands around the trees can be problematic as moisture stress can occur. The identification of a suitable less competitive ground cover species is still being sought experimentally.

With closely spaced planting designs, the trees themselves will increasingly shade out many weed species. However, total canopy closure may not be desirable as it can cause the death of lower tree branches.

ALWAYS READ THE PRODUCT LABEL

3 Plantation systems

Christmas tree plantations can be conveniently divided into two main types.

Forest plantations

In conventional forest plantations, early respacing operations may make trees available for sale as Christmas trees. Crops may also be grown specifically as Christmas trees, within a forest block, as part of normal forest production. These trees will come under the forestry definition for herbicide use – users can only apply products with approval for use in forestry situations.

On the more infertile upland sites, a pre-plant application of glyphosate followed by cultivation may give a season or more control of weeds. Thereafter emerging grass and herbaceous weeds should be controlled using conventional forestry weeding regimes. Typically this may involve a directed application of glyphosate, followed by an overall spray of propyzamide in the winter. Willoughby and Dewar (1995) give more information on forestry weed control.

The practice of this form of Christmas tree cultivation involves a relatively low input of resources, giving a lower return from usually poorer quality trees. Attempting this form of management on more fertile lowland restock sites can create problems, as prodigious weed growth can be difficult to control with approved herbicides. Later sections in this Field Book deal with the use of herbicides in horticultural situations – the techniques required for weed control on fertile plantation sites are similar, but fewer   products are available for use. This Field Book identifies which   herbicide products have full forestry approval. A full list of forestry approved products is given in the Appendix.

Horticultural/nursery plantations

Trees established on horticultural or nursery holdings, planted specifically to produce Christmas trees, are regarded as hardy ornamental nursery stock, and therefore come under the long-term off-label arrangements for the use of herbicides (detailed in the Appendix). Any herbicide product with approval for use on any growing crop may be used, as long as all the conditions of use detailed on the label are fulfilled. Applications are made entirely at the user's own commercial risk. **Christmas trees grown in a separate fenced-off area within a forest block, where there is no intention of growing trees on to form a final timber crop, could be regarded as being grown on a horticultural holding, provided there is no risk of exposure to the public, to wildlife, or non-target fauna, and there is no risk of run-off to watercourses and surrounding areas.**

Specialist growing of Christmas trees in this manner requires higher inputs, but there is also potential for greater returns through the production of top quality trees. Weed control will need to be more intensive, as weed growth is likely to be greater on the more fertile sites that will tend to be chosen for specialist growing. The remainder of this Field Book will deal specifically with weed control in horticultural/nursery production of Christmas trees.

3

4 Specialist production – weed control regimes

The aim should be to keep a site weed-free at all times, with control of grasses and creeping or climbing herbaceous weeds being the highest priority. Weeding should be seen as a preventive rather than a reactive measure – it is usually far easier to control weeds before they have become established. It is also possible to use cheap, tractor-mounted applications for many pre-emptive weed control treatments, although some directed sprays may be necessary.

A pre-plant spray of a broad spectrum herbicide should be used to kill off any established weeds prior to planting. Any cultivation operations should aim to provide an even, firm, fine tilth across the entire site, for the effective use of soil-acting residual herbicides. Soil-acting herbicides should be applied to weed-free sites in the winter or spring, as an overall application after planting. If weeds start to emerge during the growing season, they will need to be controlled with either directed or overall sprays, depending on the weeds present and hence the products used. At the end of the growing season, established weeds will need to be controlled, again either through directed or overall sprays, prior to the re-application of soil-acting herbicides in the winter/spring. Alternatively, some products exist that may control established weeds and give a degree of residual control in the second year. The post-planting cycle should be repeated for the life of the rotation.

5 Herbicide choice

Tables 1 and 2 summarize some of the most useful herbicides for use in Christmas tree weed control. Only those products with forestry approval can be used for non-specialist production within forests. Tables 3 and 4 give details of specific weed species that each product will control. Where blanks appear in the tables evidence on the specific species is not available. However, broad spectrum products in particular may control a far wider range of species. Guidance on likely weed control spectrums is given in the following sections. Descriptions of individual herbicides are given in the section on herbicide characteristics.

5

ALWAYS READ THE PRODUCT LABEL

Table 1. Herbicide crop tolerance

Active ingredient	Generally safe to spray overall as detailed in Table 2	Overall sprays possible with some care – see Table 2. Some damage is possible	Use ONLY as a directed spray – extreme care needed to avoid crop damage
Asulam		✓	
Atrazine		✓	
Clopyralid	✓	✓	
Cyanazine	✓		
Cycloxidim			✓
2,4-D			✓
2,4-D + dicamba + triclopyr			✓
Dichlobenil			✓
Diuron			✓
Fluazifop-*p*-butyl	✓		
Fluroxypyr			✓
Glufosinate ammonium			✓
Glyphosate			✓
Isoxaben	✓	✓	
Lenacil	✓		

cont.

Table 1. Herbicide crop tolerance cont.

Active ingredient	Generally safe to spray overall as detailed in Table 2	Overall sprays possible with some care – see Table 2. Some damage is possible	Use ONLY as a directed spray – extreme care needed to avoid crop damage
Mecoprop (P)			✔
Metazachlor	✔		
Napropamide	✔		
Oxadiazon			✔
Paraquat			✔
Pendimethalin	✔		
Pentanochlor			✔
Propaquizafop	✔		
Propyzamide	✔		
Simazine	✔		
Triclopyr			✔

Notes

Table 2 and the text give essential information on rates and timings for overall or directed applications.

In all circumstances, and particularly when using herbicides with foliar activity, directed sprays are the safest means of application. Any overall application has the potential to cause some distortion or necrosis of foliage. Crop tolerance in this table refers to spruce, pine, and fir genera, and not to specific Christmas tree species. Users are advised to test small areas when using products they are unfamiliar with, prior to making large-scale commercial applications.

Table 2. Herbicide options and rates

Active ingredient	Example product [5]	Formulation	Manufacturer	Pre-crop planting	Pre-crop flushing; residual control of germinating weeds on clean sites	Post-flushing control of established weeds	End-of-season clean up	Pre-crop flushing; control of established weeds with residual activity	Cost for 1 l/kg (1995) (£) [4]	Cost per hectare (£)	Forestry approval [6]
Asulam	Asulox	400 g/l	Rhône Poulenc	✓5–10 l/ha	X	✓5–10 l/ha	✓10 l/ha	X	9	45–90	✓
Atrazine	Unicrop Flowable Atrazine	500 g/l	Unicrop	X	✓3–5 l/ha	X	X	✓9 l/ha	3	9–27	✓
Clopyralid	Dow Shield	200 g/l	DowElanco	✓0.5–1 l/ha	X	✓0.5–1 l/ha	✓0.5–1 l/ha	X	65	33–65	✓
Cyanazine	Fortrol	500 g/l	Cyanamid	X	✓1.5–5 l/ha	✓1.5–5 l/ha	X	✓5 l/ha	13	13–65	X
Cycloxidim	Laser	200 g/l	BASF	✓0.75–2.25 l/ha	X	✓0.75–2.25 l/ha	✓0.75–2.25 l/ha	X	50	38–101	X
2,4-D	Dicotox Extra	400 g/l	Rhône Poulenc	✓2–8 l/ha	X	D[1] ✓2–8 l/ha	D[1] ✓2–8 l/ha	X	4	8–32	✓
2,4-D + dicamba + triclopyr	Broadshot	200:85:65 g/l	Cyanamid	✓3–5 l/ha	X	D ✓3–5 l/ha	D ✓3–5 l/ha	X	18	54–90	✓
Dichlobenil	Casoron G	6.75% w/w	Nomix Chipman	X	D✓[3] 56 kg/ha	X	X	D ✓[3] 56 kg/ha	3	168	X
Diuron	Freeway	500 g/l	Rhône Poulenc	X	D✓2.5 l/ha[8]	X	X	X	10	25	X
Fluazifop-p-butyl	Fusilade 250 EW	250 g/l	Zeneca	✓0.5–1.5 l/ha	X	✓0.5–1.5 l/ha	✓0.5–1.5 l/ha	X	76	38–114	X
Fluroxypyr	Starane	200 g/l	DowElanco	✓1 l/ha	X	D✓1 l/ha	D✓1 l/ha	X	28	28	X
Glufosinate ammonium	Challenge	150 g/l	AgrEvo	✓5 l/ha[7]	X	D✓5 l/ha	D✓5 l/ha	X	11	55	✓
Glyphosate	Roundup ProBiactive	360 g/l	Monsanto	✓5 l/ha	X	D✓1.5–5 l/ha	D[2] ✓1.5–2 l/ha	X	7	11–35	✓
Isoxaben	Flexidor 12	125 g/l	DowElanco	X	✓2 l/ha	X	X	X	51	102	✓

cont.

9

Table 2. Herbicide options and rates *cont.*

Active ingredient	Example product [5]	Formulation	Manufacturer	Pre-crop planting	Pre-crop flushing; residual control of germinating weeds on clean sites	Post-flushing control of established weeds	End-of-season clean up	Pre-crop flushing; control of established weeds with residual activity	Cost for 1 l/kg (1995) (£) [4]	Cost per hectare (£)	Forestry approval [6]
Lenacil	Venzar Flowable	440 g/l	Du Pont	X	✓5 l/ha	X	X	X	28	140	X
Mecoprop (P)	Duplosan	600 g/l	BASF	✓2.5 l/ha	X	D✓ 2.5 l/ha	D✓2.5 l/ha	X	10	25	X
Metazachlor	Butisan S	500 g/l	BASF	X	✓2.5 l/ha	X	X	X	30	75	X
Napropamide	Devrinol	450 g/l	Rhône Poulenc	X	✓2.1–9 l/ha	X	X	X	36	76–324	X
Oxadiazon	Ronstar Liquid	250 g/l	Rhône Poulenc	X	✓D 4–8 l/ha	✓ D 4–8 l/ha	X	X	25	100–200	X
Paraquat	Gramoxone 100	200 g/l	Zeneca	✓5.5 l/ha	X	D✓5.5 l/ha	D✓5.5 l/ha	X	8	44	✓
Pendimethalin	Stomp 400	400 g/l	Cyanamid	X	✓5 l/ha	X	X	X	9	45	X
Pentanochlor	Croptex Bronze	400 g/l	Hortichem	X	✓D5.6–11 l/ha	✓D5.6–11 l/ha	X	X	13	73–143	X
Propaquizafop	Falcon	100 g/l	Cyanamid	✓1.5 l/ha	X	✓1.5 l/ha	✓1.5 l/ha	X	48	72	✓
Propyzamide	Kerb Flowable	400 g/l	PBI	✓3.75 l/ha	✓3.75 l/ha	X	X	✓3.75 l/ha	24	90	✓
Simazine	Unicrop Flowable Simazine	500 g/l	Unicrop	X	✓2.2–4.5 l/ha	X	X	X	4	9–18	X
Triclopyr	Timbrel	480 g/l	DowElanco	✓4 l/ha	X	D✓4 l/ha	D✓4 l/ha	X	20	80	✓

See following page for notes to Table 2.

Notes (for Table 2)

1 – Will tolerate over-spraying provided leader growth has hardened. However, some foliage damage may occur, so it is safest to use directed sprays. Do not treat trees less than 1 m in height.

2 – Will tolerate over-spraying at 1.5 l/ha provided leader growth has hardened. Use directed sprays if possible to avoid foliage damage. Directed sprays of up to 5 l/ha may be used.

3 – Use 30–40 kg/ha for trees established less than 2 years after planting. Apply as a directed spot over weeds, around crop trees.

4 – Costs are indicative only, and do not take account of any discounts or extra costs for order size, etc. The costs per hectare shown are per treated hectare for herbicide only, and do not include the cost of application. Thus if spots or strips around trees are treated instead of the whole area, costs will be considerably less for a gross hectare of woodland.

5 – Due to the wide range available, only sample product names are given. Users should contact individual manufacturers for specific costs and conditions of use for their products. For those products that have a full forestry approval, users should refer to the Appendix for a guide to the **specific products** with the same active ingredients that are approved for use in forestry.

6 – Those products **without** forestry approval **can only** be used with Christmas trees grown in horticultural systems, under the long-term off-label arrangements.

7 – Use 3 l/ha for small weeds, and 5 l/ha for large deep-rooted species.

8 – Take particular care when using diuron, and consider alternative treatments, particularly on lighter soils, on poorly established crops and when unfamiliar with the product.

X – Unsuitable treatment.

D✓ – Directed sprays only should be used at the rate indicated.

✓ – Suitable herbicide treatment. Products can be sprayed over the crop at the rate indicated (except in pre-planting situations, when no crop is present), however directed sprays are recommended as the safest option.

11

Table 3. Susceptibility of germinating weeds to selective pre-emergent herbicides

	Atrazine	Cyanazine	Dichlobenil	Diuron	Isoxaben	Lenacil	Metamitron	Metazachlor	Napropamide	Oxadiazon	Pendimethalin	Propyzamide	Simazine
American willowherb	S†	MS†	!	S	MR†	S	MS	S	S		MS†	R	S†
Annual meadow grass	S	S	!	S	R	S	S	S	S	S	S	S	S
Awned canary grass			!								S		
Barren brome		S	!					MS				S	
Bents	S		!	R								S	
Bittersweet			!									S	
Black bindweed	MS	S	!	MS	S	S	MR	MS	MS	S	S	S	MS
Black grass	S	MS	!		S	R	MR	S	S		MS	S	S
Black mustard	S	S	!					MR				S	S
Black nightshade	S	MS	!	S		MS	MR		R		S	S	S
Bracken			!	S			S						
Broadleaved dock	S	S	!	S	S		MS	MR	R	S		S	S
Charlock	S	S	!	S	S	S	MS	MS		S	S*	S	MR
Cleavers	MR	MR	!	MR	MS	R	R	S	S†		S†	MS†	MR†
Clover (from seed)	S†	S†	!		S†		S	S†				S	
Cocksfoot	MR		!									S	
Common chickweed	S	S	!	S	S	S	S	S	S		S	S	S
Common couch	MR		!	R								S	

cont.

Table 3. Susceptibility of germinating weeds to selective pre-emergent herbicides *cont.*

	Atrazine	Cyanazine	Dichlobenil	Diuron	Isoxaben	Lenacil	Metamitron	Metazachlor	Napropamide	Oxadiazon	Pendimethalin	Propyzamide	Simazine
Common fleabane								MS					
Common fumitory	MS	MS	–	R	S	S	MS	R	S	S	S	MS	MS
Common hemp-nettle	S	S	–		S	R		MR	S		S		S
Common orache	MS	S	–	S	S	S	S	S	S	S	S		MS
Common poppy	S	S	–	S	S	S	S	S	S		S	S	S
Common speedwell	S	S	–	R	S	MR	S	S	S	S	S	S	MS
Corn buttercup	R		–		S	S	MS				S		
Corn chamomile	S		–	S	S	S	S	S	R		S		S
Corn marigold	S		–		S	S	S	S	S		S		S
Corn spurrey	MS		–	S	S		S	MS			S	S	MS
Creeping bent (watergrass)	S		–									S	
Creeping buttercup	MS†	MR†	–	MR	MS†		MS†	S†	MS†		MS†	S	MR†
Creeping soft grass	MR		–		R							S	
Creeping thistle			–	R						R			
Crested dog's tail			–									S	
Curled dock			–				S		S			S	
Cut-leaved crane's-bill			–					S				S	MR
False oat grass	MR		–									S	

cont.

13

Table 3. Susceptibility of germinating weeds to selective pre-emergent herbicides cont.

	Atrazine	Cyanazine	Dichlobenil	Diuron	Isoxaben	Lenacil	Metamitron	Metazachlor	Napropamide	Oxadiazon	Pendimethalin	Propyzamide	Simazine
Fat-hen	S	MS	–	S	S	S	S	MS	S	S	S	S	S
Fescues	MS	S	–									S	S
Field forget-me-not	S	S	–		S		S	S			S	S	S
Field gromwell			–	R				S					S
Field horsetail			–									S	
Field pansy	MS	MS	–	S	S	R	S	MR	MS		S		MS
Fool's parsley		MR	–				S						
Foxglove			–									R	
Germander speedwell	S	S	–		S	MR		S		S	S		MR
Greater plantain	S	S	–										
Green speedwell	S	S	–		S	MR		S		S	S		MR
Grey speedwell	S	S	–		S	MR		S		S	S		MR
Groundsel	S	S	–	MR	S	MS	S	S	S	S			S
Hairy bitter cress			–		S	S	S	S	MS				
Henbit dead-nettle	S	S	–	MR			MS		S	S	S		
Hoary plantain	S	S	–	R									S
Ivy-leaved speedwell	S	S	–	R	S	R	MS	S	MS		S	S	MS
Knotgrass	MR	MS	–	MS	S	S	S	R			S	S	MS
Mat grass			–										MR

cont.

Table 3. Susceptibility of germinating weeds to selective pre-emergent herbicides *cont.*

	Atrazine	Cyanazine	Dichlobenil	Diuron	Isoxaben	Lenacil	Metamitron	Metazachlor	Napropamide	Oxadiazon	Pendimethalin	Propyzamide	Simazine
Meadow fescue	MS		–									S	
Meadow foxtail			–									S	
Pale persicaria	MS	S	–		S	S	MS		MS		S		MS
Parsley piert	S	S	–		S†			S			S†		S
Perennial/stinging/common nettle	S†	S†	–	S	S	S		S†	S†		S†	S†	S†
Pineapple weed	S		–		S			S	S		S		S
Plantains	S		–	R									
Purple moor grass	R		–		MS†							S	
Ragwort	S†	MS†	–		S			S†	MS†		MR†	MR†	S†
Red dead-nettle	S	S	–	MR	S	MS	MS	S	S		S		S
Redshank	MS	S	–		S	S	MS	MS	MS	S	S	S	MS
Ribwort plantain	S		–	R	S					S			
Rosebay willowherb	MS	S	–	S	R			S	S		S	R	S
Rough meadow grass	S	S	–								S	S	MR
Rye grasses	S	S	–	S	S	S			S			S	S
Scarlet pimpernel	S	S	–		S	S	MR	S	R		S		S
Scented mayweed	S	S	–		S	S	S	S	S	S	S	S	S
Scentless mayweed	S	S	–	S	S	S	S		S	S	S	S	S

cont.

15

Table 3. Susceptibility of germinating weeds to selective pre-emergent herbicides *cont.*

	Atrazine	Cyanazine	Dichlobenil	Diuron	Isoxaben	Lenacil	Metamitron	Metazachlor	Napropamide	Oxadiazon	Pendimethalin	Propyzamide	Simazine
Sedges			–									MS	
Sheep's sorrel	MR	S	–									S	S
Shepherd's purse	S	S	–	S	S	S	S	S	S	S	S	MS	S
Small nettle	S	S	–	S	S	MS	S	MS	MS	S	S	S	MR
Smooth meadow grass	MS		–		R							S	S
Smooth sow thistle			–	MS		S			S	S			
Soft brome			–									S	
Spear thistle	S†	S†	–		MR†			S†	MR†		MR†	R†	S†
Stinking chamomile	S		–	S	S	S		S			S		S
Sweet vernal grass	S		–									S	
Thale cress	S		–					S					
Timothy			–									S	
Tufted hair grass	MR		–									S	MR
Vetches (from seed)			–										
Volunteer cereals	MS		–					MR	MS			S	
Volunteer oilseed rape			–		S			R			S*		
Wall speedwell	S	S	–		S			S		S	S	S	MS
Wavy hair grass	S		–								S	S	
White dead-nettle	S	S	–							S	S		

cont.

Table 3. Susceptibility of germinating weeds to selective pre-emergent herbicides *cont.*

	Atrazine	Cyanazine	Dichlobenil	Diuron	Isoxaben	Lenacil	Metamitron	Metazachlor	Napropamide	Oxadiazon	Pendimethalin	Propyzamide	Simazine
White mustard	S	S	!	S				MR					S
Wild carrot			!				S						
Wild oat	MS		!			R	R	MR			S		MS
Wild pansy			!					MR			S		
Wild radish	S	S	!	S	S	S	MR					S	S
Wood small reed			!							S		S	
Yellow oat grass			!						S			S	
Yorkshire fog	S		!									S	

Key

S	-	Susceptible.
MS	-	Moderately susceptible.
MR	-	Moderately resistant.
R	-	Resistant.
!	-	Not tested.
*	-	Most germinating weeds are likely to be susceptible to dichlobenil.
†	-	Plants arising from deep-germinating seeds may not be controlled.
		Information based upon glasshouse pot trials where activity is likely to be greater than in field conditions.
		These results give an indication of likely susceptibility, but should be treated with a degree of caution.

This table lists primarily those weeds detailed on product labels. Some additional information based upon Forestry Commission experimental results is included.

17

Table 4. Susceptibility of common weeds to selective post-emergent herbicides

Weed	Asulam	Atrazine	Clopyralid	Cyanazine	Cycloxidim	2,4-D	2,4-D + Dicamba + Triclopyr	Dichlobenil	Fluazifop-p-butyl	Fluroxypyr	Glufosinate ammonium	Glyphosate	Mecoprop	Metamitron	Metazachlor	Oxadiazon	Paraquat	Pentanochlor	Propaquizafop	Propyzamide †	Triclopyr
Annual meadow grass		3ETL		3ETL	R			S			S	–		MS,C	2ETL		S	MS, 3ETL	3ETL	S	
Awned canary grass											S	–								S	
Barren brome		FT		2ETL	FT				FT		S	–					S			S	
Bents									4ETL		S	–					MS				
Bittersweet												–								MS	
Black bindweed		50mm	2ETL	100mm				S		6ETL	S	–		MR			S			MS	
Black grass		FT		2ETL	FT			S	FT		S	–		MR	2ETL		S	S	FT	S	
Black mustard		100mm		6ETL				S				–						R			
Black nightshade		100mm		100mm				R			S	–		MR		S		S		MS	
Bracken	S							S			S	–		C				R			
Broadleaved dock	S						S	S		S	S	–	S							MS	
Cat's ear							S						S								
Charlock		100mm		6ETL				S		R	S	–		MS,C			S	3ETL			
Cleavers								S		S	S	–		R		S,C	S	S		R	
Clover			2ETL			S								C							

cont.

Table 4. Susceptibility of common weeds to selective post-emergent herbicides cont.

Weed	Asulam	Atrazine	Clopyralid	Cyanazine	Cycloxidim	2,4-D	2,4-D + Dicamba + Triclopyr	Dichlobenil	Fluazifop-p-butyl	Fluroxypyr	Glufosinate ammonium	Glyphosate	Mecoprop	Metamitron	Metazachlor	Oxadiazon	Paraquat	Pentanochlor	Propaquizafop	Propyzamide †	Triclopyr
Cocksfoot		MR															MS			MR	
Coltsfoot		100mm	MS, 6ETL#								S	–	S	C			S	S		S	
Common chickweed		100mm						S		S	S	–								S	
Common couch		3ETL		100mm	FT			S	4ETL		S	–			4ETL		MS		3ETL		
Common daisy						S	MS			S	S	–									
Common dandelion						S	MS				S	–									
Common fleabane										2ETL	S	–									
Common fumitory		50mm		1ETL						S	S	–		MS,C			S	S		R	
Common hemp-nettle				100mm							S	–					S	3ETL			
Common orache		50mm		1ETL				S		R	S	–		C			S	S			
Common poppy		100mm						S			S	–		C	2ETL		S				
Common speedwell		100mm		100mm				S			S	–		C			S			MS	
Corn buttercup		R					S					–		MS,C			S				
Corn chamomile		100mm						S		R	S	–						R			
Corn marigold		100mm	6ETL					S		2ETL	S	–		C	2ETL		S	R			
Corn spurrey		50mm									S	–		C			S	S		S	

cont.

19

Table 4. Susceptibility of common weeds to selective post-emergent herbicides *cont.*

cont.

Weed	Asulam	Atrazine	Clopyralid	Cyanazine	Cycloxidim	2,4-D	2,4-D + Dicamba + Triclopyr	Dichlobenil	Fluazifop-p-butyl	Fluroxypyr	Glufosinate ammonium	Glyphosate	Mecocrop	Metamitron	Metazachlor	Oxadiazon	Paraquat	Pentanochlor	Propaquizafop	Propyzamide †	Triclopyr
Creeping bent (watergrass)	S	3ETL			FT				4ETL		S	!					MS			S	
Creeping buttercup						S	R				S	!		MS,C						MS	
Creeping soft grass		MR				S					S	!					MS			MS	
Creeping thistle	S		MS, 250mm#				S				S	!								S	
Crested dog's tail											S	!								MS	
Curled dock	S					S	S			S	S	!									
Cut-leaved crane's-bill							R			R	S	!			C						
False oat grass							S			S	S	!					S	S		S	
Fat-hen		100mm		2ETL							S	!		MS,C		S	MS			MS	
Fescues		MS									S	!					S			S	
Field forget-me-not		100mm		4ETL			S			S	S	!		C			S			S	
Field horsetail										R		!			2ETL	S				MS	
Field pansy				1ETL							S	!		MS,C			S	3ETL			
Field/perennial bindweed														C							
Fool's parsley				1ETL														R			
Foxglove																				R	

Table 4. Susceptibility of common weeds to selective post-emergent herbicides cont.

Weed	Asulam	Atrazine	Clopyralid	Cyanazine	Cycloxidim	2,4-D	2,4-D + Dicamba + Triclopyr	Dichlobenil	Fluazifop-p-butyl	Fluroxypyr	Glufosinate ammonium	Glyphosate	Mecoprop	Metamitron	Metazachlor	Oxadiazon	Paraquat	Pentanochlor	Propaquizafop	Propyzamide + Triclopyr	Triclopyr
Germander speedwell		100mm		100mm				S			S	–!			2ETL		S				
Greater plantain		100mm		100mm		S	S	S			S	–!	S								
Green speedwell		100mm		100mm				S			S	–!			2ETL		S				
Grey speedwell		100mm		100mm				S			S	–!			2ETL		S				
Ground elder		100mm						MS			MS	–!									
Groundsel			6ETL	1ETL				S		MS 2ETL	S	–!		MS,C	2ETL			MS 3ETL			
Hawkweed							S	S			MS	–!	S								
Henbit dead-nettle		100mm		100mm			S	S		4ETL	MS	–!		MS,C	2ETL			3ETL			
Hoary plaintain						S	S	S			S	–!	S								
Ivy-leaved speedwell		100mm		100mm				S		MS, 2ETL 2ETL	S	–!		MS,C	2ETL		S	3ETL			
Knotgrass			MS 2ETL	1ETL			S	R			S	–!		MS,C	2ETL		S				
Meadow buttercup		MS														C				MS	
Meadow fescue	MS																			MS	
Meadow foxtail																					
Pale persicaria	50mm	100mm	MS 2ETL	2ETL						MS 2ETL	S	–!		MS,C			S	S		S	
Parsley piert	100mm			1ETL								–!	MS				S			S	

cont.

21

Table 4. Susceptibility of common weeds to selective post-emergent herbicides cont.

Weed	Asulam	Atrazine	Clopyralid	Cyanazine	Cycloxidim	2,4-D	2,4-D + Dicamba + Triclopyr	Dichlobenil	Fluazifop-p-butyl	Fluroxypyr	Glufosinate ammonium	Glyphosate	Mecocrop	Metamitron	Metazachlor	Oxadiazon	Paraquat	Pentanochlor	Propaquizafop	Propyzamide + Triclopyr	Triclopyr
Pearlwort						S															S
Perennial sow thistle			MS, 250mm#								S	!								S	
Perennial/stinging/common nettle											S	!									
Pineapple weed		100mm	6ETL							S	S	!			4ETL			R			
Purple moor grass		R						S				!									
Ragwort		100mm	MS				MS	S				!									
Red dead-nettle		100mm		100mm			S			4ETL	S	!		MS,C	2ETL			3ETL		MS	
Redshank		50mm	MS, 2ETL	100mm			S			MS, 2ETL	S	!		MS,C			S	S			
Ribwort plantain						S	S					!	S								
Rosebay willowherb		50mm		FT	MR				FT			!					MS			R	
Rough meadow grass		S									S	!		MR			MS			S	
Rye grasses		3ETL		3ETL						R	S	!					MS			S	
Scarlet pimpernel		100mm		100mm							S	!						3ETL			
Scented mayweed		100mm	6ETL	2ETL						MS, 2ETL	S	!		C	4ETL		S	R			
Scentless mayweed		100mm	6ETL	2ETL				S		MS, 2ETL	S	!		C	4ETL		S	R			
Sedges												!								MS	

Table 4. Susceptibility of common weeds to selective post-emergent herbicides cont.

Weed	Asulam	Atrazine	Clopyralid	Cyanazine	Cycloxidim	2,4-D	2,4-D + Dicamba + Triclopyr	Dichlobenil	Fluazifop-p-butyl	Fluroxypyr	Glufosinate ammonium	Glyphosate	Mecoprop	Metamitron	Metazachlor	Oxadiazon	Paraquat	Pentanochlor	Propaquizafop	Propyzamide †	Triclopyr
Self-heal						S							S								
Sheep's sorrel		50mm										—								MS	
Shepherd's purse		100mm		100mm		S	S			R	S	—		C			S	3ETL		R	S
Small nettle		MS		100mm			S			R	S	—		C			S	3ETL		MS	
Smooth meadow grass												—								S	
Smooth sow thistle			6ETL								S	—									
Soft brome											S	—									
Spear thistle												—				S		S		S	
Stinking chamomile		100mm	MS, 250mm# 2ETL			S	S				S	—	S							S	
Sweet vernal grass		S										—						R			
Timothy									FT			—					MS				
Trefoils		MR	2ETL									—									
Tufted hair grass					FT						MS	—					MS			S	
Vetches			2ETL									—							FT	S	
Volunteer cereals		FT									S	—					S			S	
Volunteer oilseed rape										R	S	—			2ET		S			S	
Wall speedwell		100mm		100mm			S				S	—									

cont.

23

Table 4. Susceptibility of common weeds to selective post-emergent herbicides cont.

	Asulam	Atrazine	Propyzalid	Cyanazine	Cycloxidim	2,4-D	2,4-D + Dicamba + Triclopyr	Dichlobenil	Fluazifop-P-butyl	Fluroxypyr	Glufosinate ammonium	Glyphosate	Mecoprop	Metamitron	Metazachlor	Oxadiazon	Paraquat	Pentanochlor	Propaquizafop	Propyzamide †	Triclopyr
Wavy hair grass		S							MS		MS	!									
White dead-nettle		100mm		100mm	FT						S	!									
White mustard		100mm		6ETL				S				!		C			S	3ETL		S	
Wild carrot			2ETL									!							FT		
Wild oat		FT			FT			S	FT		S	!		R			S	R			
Wild pansy											S	!									
Wild radish		50mm		100mm						R	S	!		MS,C				S		S	
Wood small reed												!					MS			S	
Yellow oat grass																				S	
Yorkshire fog		S										!									

See following page for key to Table 4.

Key (for Table 4)

Post-emergent growth stage of weeds (latest at which controlled):

C	—	Cotyledon.
ETL	—	Number of expanded true leaves.
mm	—	Diameter or height of weeds.
FT	—	Fully tillered.
S	—	Susceptible at all growth stages.
MS	—	Moderately susceptible at all growth stages.
MR	—	Moderately resistant at all growth stages.
R	—	Resistant.
	—	Not tested.
†	—	All weed susceptible for propyzamide post-emergence are for fully established weeds.
#	—	Control with a programme of an application at 0.5 l/ha followed by one of 1.0 l/ha 3–4 weeks later.
!	—	No specific weeds are listed for glyphosate. It is likely that most species will be controlled by applications of glyphosate at 5.0 l/ha provided they are actively growing.

This table lists only those weed species identified on product labels. It is likely that a wider range of species will be controlled than those listed in the table. No guarantees are made by manufacturers, but the following are likely:

• In addition to the weeds listed, most species will be suppressed by applications of glufosinate ammonium at 5.0 l/ha, or paraquat at 5.5 l/ha, providing they are actively growing, although repeated applications may be required to achieve a total kill of deep-rooted species;

• 2,4-D + dicamba + triclopyr, 2,4-D, and triclopyr will all control many herbaceous species.

Pre-planting

Sites may be cleared of perennial weeds by controlling them in a previous crop, or leaving a site fallow then spraying out the weeds when they re-grow. Trees may be planted into bare ground following the desiccation of grass or a crop, or into a cultivated soil.

Herbicide options

Products and product mixes based upon:

2,4-D
2,4-D + dicamba + triclopyr
Fluroxypyr
Glufosinate ammonium
Glyphosate
Mecoprop (P)
Paraquat
Triclopyr

Pre-flushing residual control on clean sites

Residual soil-acting herbicides should be used to prevent the germination of weed species during the growing season. Application of soil-acting residual herbicides is generally more effective than treating weeds when they emerge. Some residual products offer good crop tolerance and can be sprayed over dormant trees. These herbicides should be applied to clean sites, preferably free of brash and detritus, and which, for the first application after planting, have been cultivated to a firm, fine tilth. Rainfall is required after application to incorporate the herbicides into the soil. A tank mix of products is often used – refer to the weed susceptibility Tables 3 and 4 to pick complementary products covering a wide spectrum of weeds. Aim to apply in spring, **before** bud-burst and weed emergence.

Herbicide options

Products and product mixes based upon:

Atrazine
Cyanazine
Dicamba

ALWAYS READ THE PRODUCT LABEL

Dichlobenil*
Isoxaben
Lenacil
Metazachlor
Napropamide
Oxadiazon
Pendimethalin
Propyzamide
Simazine

* Not in first year after planting.

Post-flushing control of established weeds and end-of-season clean up

Even with effective applications of soil-acting residual herbicides in the winter and spring, some control of emerging weeds during the growing season may be required. Although some selective products offer fairly good crop tolerance, most broad spectrum herbicides must be applied using directed sprays, to prevent damage to the crop.

Towards the end of the growing season, established weeds should be controlled prior to the application of soil-acting herbicides in the subsequent winter/spring. Some selective products only control weeds of a certain size (see Table 4), and may not therefore be effective at the end of the growing season. Glyphosate may be sprayed over healthy fully dormant trees of most Christmas tree species, but it may be safer to use directed sprays – this also allows the use of higher product rates to control deep-rooted weed species. Weeds must be green and actively growing at the time of spraying, but it should be recognized that control may take many weeks in colder weather conditions. Never add any additive or wetter to glyphosate when making over the top sprays. Triclopyr and 2,4-D can be used over dormant trees of some species to control perennial broadleaved weeds, although again a directed spray would be a safer option. The other non-selective products listed here – paraquat, glufosinate ammonium, and 2,4-D + dicamba + triclopyr – should only be used as directed sprays.

5

Herbicide options

Products and product mixes based upon:

Asulam
Clopyralid
Cycloxidim
2,4-D
2,4-D + dicamba + triclopyr
Fluazifop-p-*butyl*
Fluroxypyr
Glufosinate ammonium
Glyphosate
Mecoprop (P)
Pentanochlor
Propaquizafop
Triclopyr

Pre-flushing control of established weeds with residual activity

Most sites are cleaned up with an application of foliar-acting herbicide at the end of the growing season, prior to the application of soil-acting herbicides to the bare site in the spring. An alternative is to apply products that will control both established weeds and give residual control into the following growing season.

Herbicide options

Dichlobenil
Propyzamide

ALWAYS READ THE PRODUCT LABEL

6 Herbicide characteristics (for all situations)

Only herbicides with a high level of crop tolerance should be used over poorly planted or stressed trees. Trees are likely to be more susceptible to herbicides in their first growing season. In all situations directed sprays present the safest option. In general, spruces are likely to be more tolerant to herbicide applications than pines, which are in turn likely to be more tolerant than firs.

Asulam

Asulam is a foliar-acting translocated herbicide that is effective on bracken and docks.

Crop trees are tolerant of over-spraying, although they may suffer some chlorosis and check in growth if treated when actively growing.

Atrazine

Atrazine is a soil-acting residual herbicide, used for the control of germinating annual grasses and broadleaved weeds. At the rates listed in this Field Book, atrazine will have little effect on perennial species. Atrazine has no soil activity in soils with an organic peat layer.

Spraying must take place before tree bud-swell in the spring. Crop trees are tolerant of overall sprays at this stage. Planting can be carried out immediately after treatment but the level of weed control will be reduced if the soil is badly disturbed after spraying.

Clopyralid

A foliar-acting herbicide for the control of some mayweed, groundsel and thistles. Spruces and firs will tolerate over-spraying, particularly when young growth has hardened off. Slight twisting of foliage may occur, but this is normally outgrown.

Cyanazine

A foliar- and soil-acting herbicide for the control of germinating grass and herbaceous weeds. Small established weeds may be controlled as well as giving some residual control of germinating weeds. Crop trees are tolerant of overall sprays when dormant.

Cycloxidim

A foliar-acting translocated herbicide for the control of annual and perennial grasses. Cycloxidim will have no effect on herbaceous species, and crop tolerance is good.

2,4-D

2,4-D is a plant growth regulating herbicide to which many herbaceous species are susceptible. Grasses are unaffected. It is absorbed mainly through the aerial parts of the plant. One week should elapse between treatment and subsequent planting. Norway spruce is moderately tolerant of overall spray providing leader growth has hardened. Other species are less tolerant. Use directed sprays whenever possible.

2,4-D + Dicamba + Triclopyr

A translocated contact herbicide which controls a wide range of annual and perennial herbaceous and woody weeds, but not grasses. Allow 1 month before replanting. All contact with crop trees must be avoided.

6

Dichlobenil

A granular herbicide for use as a spot treatment for perennial broadleaved weeds, particularly docks, nettles, and rosebay willowherb. Grasses will also be controlled. Crop tolerance is variable – use carefully directed applications paying close attention to application rates.

Diuron

A highly soluble herbicide for the control of annual grass and herbaceous weeds within established crops. Diuron has been used safely as a directed spray at rates of 2.5 l/ha. Do not use on light

ALWAYS READ THE PRODUCT LABEL

textured soils (less than 40% clay). If users are unfamiliar with this product, they should contact a professional adviser before making any applications.

Fluazifop-*p*-butyl

A foliar-acting translocated herbicide for the control of annual and perennial grasses. Fluazifop-*p*-butyl will have no effect on herbaceous species, and crop tolerance is good. It may only be applied through tractor-mounted applicators.

Fluroxypyr

A foliar-acting herbicide with good activity on annual broadleaved weeds including cleavers. Some herbaceous weeds including docks are also controlled. Crop trees will be damaged on contact – always use directed sprays.

Glufosinate ammonium

A non-translocated, foliar-acting herbicide. Glufosinate ammonium controls a broad spectrum of annual and perennial grasses and herbaceous weeds. Gradual chlorosis over the first 2–3 days after spraying is followed by a withering of treated foliage over the next 10–14 days. No translocation to rhizomes or stolons occurs – deep-rooted species may require repeat applications for a complete kill. Activity is greatest under warm, moist conditions, when weeds are actively growing. Breakdown of the product is rapid upon contact with the soil. All contact with crop trees must be avoided – always use directed sprays.

6

Glyphosate

A translocated herbicide taken up by the foliage and conveyed to the roots. It causes chlorosis and eventual death of the leaves and then kills roots and shoots. The symptoms usually become apparent about 7 days after treatment, but may take longer to show if vegetation growth is slow.

Glyphosate controls a wide range of weeds including grasses, herbaceous weeds, bracken, heather, and woody weeds. Control of some annual herbaceous weeds, such as cleavers and willowherb, may be variable.

On contact with the soil glyphosate is quickly inactivated. Planting can be carried out 7 days after treatment and a minimum of 7 days should be allowed prior to cultivation breaking up the rhizomes and roots.

Spruce, pine and Caucasian fir (Douglas fir is less tolerant) will tolerate overall sprays of 1.5 l/ha provided leader growth has hardened. Applications should be delayed until October when cold weather is likely to have hardened the crop. Hardening can occur as early as the end of July, or may be delayed until the end of October in some locations and seasons. Contact with crop species in the growing season must be avoided. Directed sprays are safer and allow higher product rates to be used to control deep-rooted weed species.

Isoxaben

A soil-acting residual herbicide giving control of germinating herbaceous weeds. Isoxaben will have no effect on grasses or established weeds. Crop tolerance is good.

Lenacil

A residual herbicide giving control of a range of germinating herbaceous and grass weed species. Lenacil will have no effect on established weeds. Crop tolerance is good.

Mecoprop/Mecoprop (P)

A foliar-acting herbicide for the control of annual herbaceous weeds such as chickweed and cleaver. Perennial herbaceous weeds such as docks, thistles, and nettles will also be controlled or suppressed. Contact with crop foliage should be avoided – always use directed sprays.

Metazachlor

A residual herbicide for the control of annual grass and herbaceous weeds. Most activity is pre-emergent, although some post-emergent weed control may be possible depending on the weed species. Crop tolerance is good, although some foliar damage may occur if sprayed on to trees in active growth.

Napropamide

A residual soil-acting herbicide for the control of annual grass and herbaceous weeds. Napropamide will have no effect on established weeds. Apply between November and the end of February. Crop tolerance is good.

Oxadiazon

A contact and residual herbicide which is particularly effective on perennial bindweed. Crop tree needles are likely to be scorched on contact, particularly after flushing – always use directed sprays.

Paraquat

Paraquat is a non-translocated contact herbicide that gives rapid control of a wide range of grass and herbaceous species, and is particularly effective on annual weeds. Deep-rooted species may require repeated applications to give effective control. All contact with crop trees must be avoided – always use directed sprays.

Paraquat is a poison, and can kill if swallowed. Glyphosate and glufosinate ammonium are safer alternative herbicides, however they may be more expensive.

Pendimethalin

A residual soil-acting herbicide for the control of annual grasses and annual herbaceous weeds. Pendimethalin will have no effect on established weeds. Crop tolerance is good.

Pentanochlor

A foliar- and soil-acting herbicide with some residual activity for the control of some annual grass and herbaceous weeds. Avoid contact with crop – always use directed sprays.

Propaquizafop

A foliar-acting translocated herbicide for the control of annual and perennial grasses. Propaquizafop will have no effect on herbaceous species. Crop tolerance is good.

6

Propyzamide

A winter-applied soil-acting herbicide which works most effectively when applied to cold, wet soils. Mild, dry weather can reduce effectiveness. Most grasses, and some herbaceous broadleaved weeds, are susceptible from germination to the true leaf stage, but herbaceous broadleaved weeds which emerge late in the season will only be partially controlled. Established grasses are also controlled. Propyzamide slowly breaks down in the soil, lasting for 3–6 months. Apply liquid formulations from 10 October until 31 December (to 31 January north of a line from Aberystwyth to London), and granular formulations from 10 October until 31 January (28 February in the north). Crop tolerance is good.

Simazine

Simazine is a soil-acting residual herbicide, used for the control of germinating annual grasses and broadleaved weeds.

Spraying must take place before tree bud-swell in the spring. Planting can be carried out immediately after treatment but the level of weed control will be reduced if the soil is badly disturbed after spraying.

Triclopyr

A plant growth regulating herbicide which is rapidly absorbed, mainly through the foliage, but also by roots and stems. Once inside the plant it is readily translocated. Triclopyr is effective against woody weeds and many perennial herbaceous weeds but not grasses.

Grasses sometimes show some yellowing following spring operations but this is quickly outgrown.

In the soil triclopyr is broken down fairly rapidly by microbial action. Planting should be deferred for at least 6 weeks after application.

Norway spruce may tolerate overall sprays up to 4 l/ha in September through to October provided leader growth has hardened. Aim to use directed sprays if possible.

ALWAYS READ THE PRODUCT LABEL

7 References

DEBOYS, R. (1994). *Mulching trial – evaluation of currently available mulching systems.* Forestry Commission Technical Development Branch Report 3/94. Forestry Commission, Edinburgh.

WILLOUGHBY, I. and DEWAR, J. (1995). *The use of herbicides in the forest.* Forestry Commission Field Book 8. HMSO, London.

7

8 Appendix

Long-term off-label arrangements

All applications made under any off-label arrangements are at the user's own risk. This means that the manufacturers cannot be held responsible for any adverse effects on crops or failure to control weeds, but employers and operators still have the responsibility to adhere to on-label instructions when using the product.

In addition to specific off-label approvals, certain fields of use are covered by long-term off-label arrangements, the revised versions of which are valid until 31 December 1999.

- All pesticides with full or provisional label approval for use on any growing crop may be used within forest nurseries, on crops prior to final planting out.

- Christmas trees grown on commercial agricultural and horticultural holdings and in forest nurseries can be regarded as hardy ornamental or forest nursery stock, and hence are covered by the same arrangements.

As well as the usual good working practices required of users, the following **additional** conditions must be complied with when applying pesticides under the long-term off-label arrangements.

- All precautions and statutory conditions of use, which are identified on the product label, must be observed.

- The method of application used must be the same as that listed on the product label, and must comply with relevant codes of practice and requirements under COSHH (The Control of Substances Hazardous to Health Regulations).

- All reasonable precautions must be taken to safeguard wildlife and the environment.

- Products must not be used in or near water unless the label specifically allows such use.

ALWAYS READ THE PRODUCT LABEL

- Aerial applications are not permitted.

- Products approved for use under protection, i.e. under polythene tunnels or glasshouses, cannot be used outside.

- Rodenticides and other vertebrate control agents are not included in these arrangements.

- Use is not permitted on land not intended for cropping – for example, paths, roads, around buildings, wild mountainous areas or nature reserves.

- Pesticides classified as hazardous to bees must not be applied when crops or weeds are flowering.

- These extensions of use apply only to label recommendations – no extrapolations are permitted from specific off-label approvals.

- Unless specifically permitted on the product label, hand-held applications are not permitted if the product label:

  (i) prohibits hand-held use;

  (ii) requires the use of personal protective clothing when using the pesticide at recommended volume rates;

  (iii) is classified as 'corrosive', 'very toxic', or has a 'risk of serious damage to eyes'.

If none of the above applies, hand-held application is permitted provided that:

  (i) the concentration of the spray volume does not exceed the maximum recommended on the label;

  (ii) spray quality is at least as coarse as the British Crop Protection Council medium or coarse spray;

  (iii) operators wear a protective coverall, boots, and gloves for applications below waist height – a face shield should be worn in addition for applications above waist height;

  (iv) if the product label gives a buffer zone for vehicle mounted use, a buffer zone of 2 m should be used for hand-held applications.

8

Approved products as of August 1996

Products with full forestry label approval

Active ingredient	Product (manufacturer)
Ammonium sulphamate	Amcide (BH&B) Root-out (Dax Products)
Asulam	Asulox (RP Environmental)
Atrazine	Atlas atrazine (Atlas) Unicrop Flowable Atrazine (Unicrop) *Note*: stocks of Unicrop Flowable Atrazine in existence at time of writing, but manufacture has ceased
2,4-D	Dicotox Extra (RP Environmental) MSS 2,4-D Ester (Mirfield)
2,4-D + dicamba + triclopyr	Broadshot (Cyanamid)
Dalapon + dichlobenil	Fydulan G (Nomix-Chipman) *Note*: stocks in existence at time of writing, but manufacture has ceased
Dicamba	Tracker (PBI)
Diquat + paraquat	Farmon PDQ (Farm Protection) Parable (Zeneca)
Fosamine ammonium	Krenite (Du Pont) *Note*: stocks in existence at time of writing, but manufacture has ceased

8

ALWAYS READ THE PRODUCT LABEL

Active ingredient	*Product (manufacturer)*
Glufosinate ammonium	Challenge (AgrEvo) Challenge 2 (AgrEvo) Harvest (AgrEvo) Headland Sword (Headland)
Glyphosate	Barbarian (Barclay) Barclay Gallup Amenity (Barclay) Danagri Glyphosate 360 (Danagri) Clayton Glyphosate (Clayton) Clayton Swath (Clayton) Glyfonex (Danagri) Glyfos (Cheminova) Glypher (PBI) Glyphogan (PBI) Glyphos Proactive (Nomix- Chipman) Glyphosate 360 (Top Farm) Hilite (Nomix-Chipman) – CDA formulation Helosate (Helm) Monty (Monsanto) MSS Glyfield (Mirfield) Roundup (Monsanto) Roundup (AgrEvo) Roundup Biactive (Monsanto) Roundup Biactive Dry (Monsanto) Roundup Pro Biactive (Monsanto) Stacato (Unicrop) Stefes Glyphosate (Stefes) Stefes Kickdown 2 (Stefes) Stetson (Monsanto) Stirrup (Nomix-Chipman) – CDA formulation Typhoon 360 (Chiltern) *(cont. on next page)*

8

Active ingredient	Product (manufacturer)
	Note: all of the glyphosate products listed have full approval, but some are not actively marketed
Imazapyr	Arsenal 50F (Nomix-Chipman)
Isoxaben	Gallery 125 (DowElanco) Flexidor 125
Paraquat	Barclay Total (Barclay) Gramoxone 100 (Zeneca/AgrEvo) Scythe LC (Cyanamid)
Propyzamide	Headland Judo (Headland) Kerb Flo (PBI, Rohm + Haas) Kerb 50W (PBI, Rohm + Haas) Kerb Granules (PBI, Rohm + Haas) Stefes Pride Flo (Stefes)
Triclopyr	Garlon 4 (DowElanco) Timbrel (DowElanco) Chipman Garlon 4 (Nomix-Chipman)

Products with full farm forestry label approval

Active ingredient	Product (manufacturer)
Fluazifop-*p*-butyl	Fusilade 5 (Zeneca) Fusilade 250 EW (Zeneca)
Propaquizafop	Falcon 100 (Cyanamid) Shogun 100 EC (Ciba-Geigy)

8

ALWAYS READ THE PRODUCT LABEL

Products with forestry off-label approval

Active ingredient	Product (manufacturer)
Clopyralid	Dow Shield (DowElanco)
Cyanazine	Fortrol (Cyanamid)
Cycloxidim	Laser (BASF)

Products with farm forestry off-label approval

Active ingredient	Product (manufacturer)
Amitrole	Weedazol (Bayer)
Cyanazine	Fortrol (Cyanamid)
Metazachlor	Butisan S (BASF)
Pendimethalin	Stomp (Cyanamid)

Full details of the method of use for products containing these active ingredients can be found in Forestry Commission Field Book 8 (Willoughby and Dewar, 1995).

8

Printed in the United Kingdom for the Stationery Office
N21289 C15 6/97 (9091)